YOUR KNOWLEDGE HAS VALUE

- We will publish your bachelor's and
  master's thesis, essays and papers

- Your own eBook and book -
  sold worldwide in all relevant shops

- Earn money with each sale

Upload your text at www.GRIN.com
and publish for free

The fundamentals of national park tourism and the potential as an economic engine

Talisa Gassmann

Bibliographic information published by the German National Library:

The German National Library lists this publication in the National Bibliography; detailed bibliographic data are available on the Internet at http://dnb.dnb.de.

ISBN: 9783389021644
This book is also available as an ebook.

© GRIN Publishing GmbH
Trappentreustraße 1
80339 München

Print and binding: Books on Demand GmbH, Norderstedt, Germany
Printed on acid-free paper from responsible sources.

The present work has been carefully prepared. Nevertheless, authors and publishers do not incur liability for the correctness of information, notes, links and advice as well as any printing errors.

GRIN web shop: https://www.grin.com/document/1442080

The fundamentals of national park tourism and the potential as an economic engine

prepared at the Harz
University

Department of Economics Course:

International Tourism Studies

Table of contents

List of illustrations

Table directory

1. Introduction

Today's world of tourism is constantly changing, characterized by increasing competition for travelers. Attracting and retaining visitors is becoming a key challenge for destinations in this dynamic environment. Against this background, sustainable tourism development has become a crucial tool for increasing the attractiveness of travel destinations. Research into the effects of tourism on the economy and quality of life of residents is essential, particularly in sensitive areas such as national parks. The challenge of designing tourism in such a way that it brings both economic benefits and optimizes the quality of life is therefore particularly demanding.[1]

As one of the most visited protected areas in the United States, Joshua Tree National Park recognizes that promoting sustainable tourism can also help increase local economic performance and quality of life. [2]That is why the question of the impact of tourism on the local and regional economy, as well as on the quality of life, is becoming increasingly important.

Despite its importance, however, there is relatively much scope for research into the impacts of tourism in Joshua Tree National Park: particularly the associated impacts and perceptions of local populations. Consequently, this bachelor's thesis aims to fill this gap by conducting a detailed case study of tourism in Joshua Tree National Park. Not only are the visit numbers and trends analyzed, but particularly relevant is the investigation and evaluation of the economic impact of tourism in the national park in order to discuss possible implications for politics and park management.

The knowledge gained serves as a basis for deriving suitable measures with which tourism in Joshua Tree National Park can be made sustainable with regard to the effects on the local population. In order to achieve this goal, a theoretical framework is first presented in the course of the work, which creates important foundations for the economic potential of tourism - as well as the associated relevance of national parks.

After explaining the research question and methodology, the case study is carried out in the main part. This includes consideration of the profile of Joshua Tree National Park , as well as current developments in the field of tourism. The effects of tourism on the economy and quality of life are then derived, with a critical analysis taking place on a positive and negative level. A solid foundation for this is provided by combining qualitative data from interviews with residents conducted by local newspaper companies and extensive quantitative research.

---

[1]See (Budeanu, 2005, p.89ff.)
[2]See (NPS.gov, 2021)

The analysis of the data collected is intended to provide concrete insights into the most relevant impacts of tourism. Based on the results, recommendations for action are derived with which tourism in Joshua Tree National Park can be made sustainable and how possible risks (e.g. increasing financial burden on the local population or a lack of living space) can be avoided.

Furthermore, the limitations of this bachelor's thesis are pointed out before the conclusions of the case study examined are summarized in a conclusion in the last chapter.

2. Theoretical background

This chapter is therefore dedicated to examining two central aspects that are crucial for understanding the research question: the economic importance of tourism and the specific role that national parks play in this context.

First, the general economic importance of tourism is discussed. As one of the largest global economic sectors, tourism has far-reaching impacts at various levels, from local communities to the global economy. [3]Following on from this, the special role of national parks in the tourism sector is examined. These unique places attract visitors from all over the world and have the potential to act as significant economic drivers. The extent to which this is possible and which economic sectors are affected is examined in more detail in the final part of the chapter. Overall, this chapter aims to provide a framework for understanding the complex interactions between tourism, national parks, local economies and populations.

2.1 The economic importance of tourism

The tourism industry was one of the fastest-expanding economic sectors before the coronavirus and continues to maintain its status as the world's dominant service industry. "(…) in 2019 alone, over 1.5 billion tourist arrivals were registered (…) with a growing importance of domestic tourism." [4]The interdependence between the increase in travel and the economic importance of tourism has been proven. Looking at the situation before the virus, it can be seen that the tourism industry contributed over 10% to global economic output. Furthermore, over 230 million jobs depend on tourism, which accounts for around 9% of global employment. [5]As a result, it is no secret that the tourism industry is a highly profitable source of income and has the ability to combat poverty and stimulate the regional economy. The latter will be examined in more detail in the next section in order to create [6]essential foundations for the further course of the work. [7] [8] [9] [10] [11]

Since the focus of this work is on the United States, the aim is to provide a brief insight into the American tourism industry. When evaluating the overall economic production value of the tourism sector, the World Travel & Tourism Council was able to determine

---

[3]See (Kumar, Hussain, 2014)
[4](BMZ, 2022)
[5]See ibid.
[6]See (Schmude, Bartl, 2023)
[7]See (BMZ, 2022)
[8]See (Adjouri, Büttner, 2008)
[9]See (Kumar, Hussain, 2014)
[10]See (Zaei, Zaei, 2013)
[11]See (German Society for International Cooperation (GIZ) GmbH, 2020)

that the USA occupied first place by a significant margin in 2007 and was able to maintain this position 10 years later with an increase of 44.94 [12]% . But when it comes to travel spending, Americans also took first place in 2005 with a total of around 82 billion dollars, and they took relatively significantly more trips within their home country.[13] [14]The reason for this lies primarily in the richness of the diverse climate zones and landscapes. This finding is particularly interesting and creates an essential knowledge base for the more detailed examination of the economic impact of tourism in Joshua Tree National Park in Chapter 5.3.[15] [16] [17] [18]

Due to various reasons, the tourism industry can make a fundamental contribution to regional development and the economy and acts as a valuable instrument. Since tourism has the characteristics of a cross-sectional industry, characterized by small and medium-sized companies, not only the hospitality industry, but also related economic sectors make valuable profits. These include, for example, various leisure providers, retailers, but also craft and agricultural businesses. Accordingly, "(…) there is great potential for increasing regional value creation by integrating regional value creation systems into the tourism value creation system." [19]In addition, there are few barriers to entry. This means that new tourism-related companies can take advantage of good starting conditions and, as a result, it is easier for them to start up. Furthermore, investments through tax revenues stimulate the regional infrastructure, which leads to the expansion of the tourist offer. If the quality of the offer is promoted, this can have a particular impact on local employment relationships and population groups with differentiated qualifications can be offered a more extensive range of opportunities. The following impact structure clearly summarizes the different contributions to regional development[20] [21] [22] [23]

---

[12]The difference of 1,378,668 corresponds to 44.94%
[13] See (Adjouri, Büttner, 2008)
[14]See (Ennew, 2014)
[15]See ibid.
[16]See (Schmude, Bartl, 2023)
[17]See (Vellas, 2011)
[18]See (Ennew, 2014)
[19](German Society for International Cooperation (GIZ) GmbH, 2020, p. 21)
[20]See (German Society for International Cooperation (GIZ) GmbH, 2020)
[21]See (Zaei, Zaei, 2013)
[22]See (Adjouri, Büttner, 2008)
[23]See (Vellas, 2011)

Figure 1: The tourism potential for regional economic development

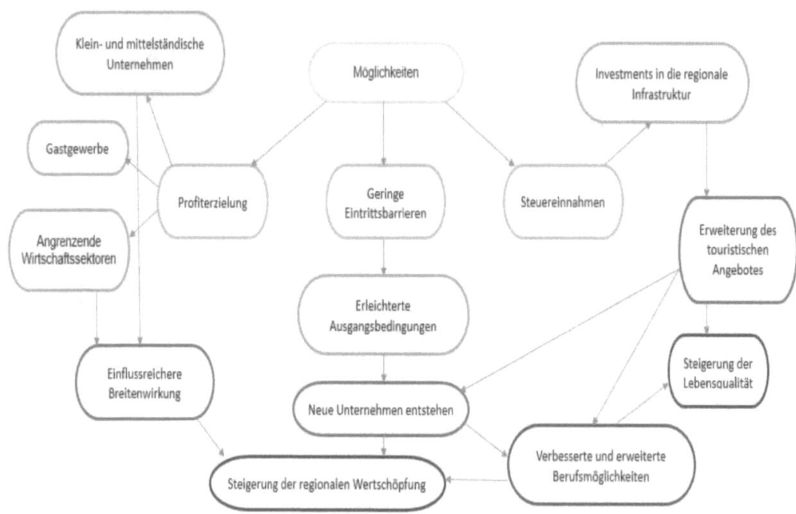

Source: own representation based on (German Society for International Cooperation (GIZ) GmbH, 2020)

Through interrelationships between various business and tourism companies, the local population, and the government, the economic impacts are a significant indicator of economic progress and planning at the regional, state and local levels. Relevant decisions in marketing and management are directly influenced. For this reason, it is extremely important to gain a deeper understanding of its importance in order to promote regional economic activity. However, due to the predominantly positive view of the economic effects, certain aspects are deliberately left out of mention. But what does that mean exactly? The inevitable chain reaction can be well explained by the infrastructure effects. On the one hand, numerous positive findings can be derived for the economy, which have consequences for the residents. The infrastructural expansion inevitably increases the attractiveness of the town. This not only brings more travelers, but also offers the population greater leisure benefits, improved connections to public transport, increased quality of entertainment and dining options, as well as various opportunities for shopping in supermarkets and boutiques. The vital basic capacity utilization, such as the range of pharmacies and doctors, is also maintained. This indicates an improved quality of life. On the other hand, this also has immediate consequences that cannot always be identified at first glance. By enhancing the image, there is greater scope for prices, which often results in a price increase for cafés, bars and restaurants. But the same trend can also be observed when it comes to rental prices and living costs. As a place becomes more valued as a vacation destination, prices can rise, thereby making more money. However, this can become an increasing financial challenge for certain population groups. In the worst case scenario, this results in part of the local population having to leave their neighborhood, leading to the phenomenon of so-called "gentrification". The consequence is that an increasing number of people with higher incomes characterize the region, with the direct change of the social structure in the community. .[24] [25] [26] [27]

Based on these findings, in the further course of the work, especially in the second half of the case study, it is primarily important to always examine both sides in terms of the effects. Now that the economic importance of tourism has been discussed, the focus should now be directed to a specific area of tourism: the national parks.

---

[24]See (Stynes, 1997)
[25]See (Schmude, Bartl, 2023)
[26]See (BMZ, 2022)
[27]See (Ennew, 2014)

2.2 The connection between national parks and tourism

2.2.1 Meaning and function

A national park not only serves the function of protecting natural heritage, but has also become an increasingly relevant tourism magnet over the decades. On the one hand, the breathtaking wilderness with its magnificent landscapes not only serves as a culture, but also as a safe haven for excursion destinations seeking relaxation. As a result, this chapter deals in more detail with the tourism significance of the national park.[28] [29]

It's hardly surprising that people in the United States have a stronger connection to national parks, since that's where the concept of wildlife refuges originated. This is therefore the main reason why the Joshua Tree National Park in California will be examined in more detail in the course of the work. The importance of a national park is of great importance for the region, surrounding communities and commercial companies. Nature reserves open up in competition as a unique destination for unmistakable positioning of relevant target groups. But how did this development of the tourism product actually come about?[30] [31] [32] [33]

The meaning and goals of tourism have constantly developed and changed over the years. "(…) [Over the course of the] [70]s, alternatives to (…) conventional forms of tourism were discussed." [34]Due to the increasing relevance of sustainability, the aspect of gentle travel became more and more important. Instead of focusing on mass tourism, sights and a fixed program, the focus was now on environmentally conscious trips with yourself, family and friends to admire the landscape. The aim was to enjoy enjoyment, activity, experiencing a cultural specialty, contact with the local population and exploring the local style. No environment could fulfill these conditions better than a national park. Values of ecotourism, sustainable tourism and alternative tourism come together there. If you look at the results of a study commissioned by the BMWA [35], the current needs coincide with the unique selling point of national parks. The results are summarized in the figure below. [36] [37] [38]

---

[28]See (National Parks Association of NSW, 2023)
[29]See (Revermann, Petermann, 2002)
[30]See (Tschurtschenthaler, 2000)
[31]See ibid.
[32]See (Klapf, 2005)
[33]See (Berzina, Livina, 2008)
[34](Klapf, 2005, p.57)
[35]Federal Ministry of Economics and Labor
[36]See (Klapf, 2005)
[37]See (National Parks Association of NSW, 2023)
[38]See (Berzina, Livina, 2008)

Figure 2: Combining the unique selling points of national parks with tourist needs

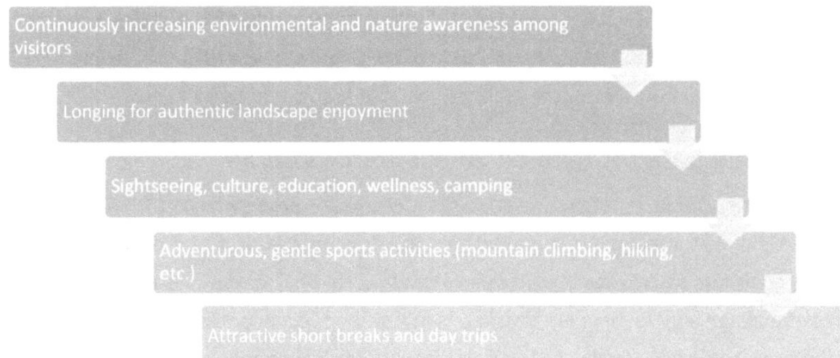

Source: own illustration based on (Klapf, 2005, p.68)

"Nature-based tourism activities are one of the fastest growing tourism market segments (…) covering 7% of all international arrivals."[39] As a result, vital interactions take place because the gastronomy, hotel industry, tour operators and local workplaces are directly connected to one another. Economic growth is triggered. Over time, the national park can create new stimuli for experiencing nature, which in turn brings with it new target groups. Ultimately, residents of the region can be enabled to perceive their landscape from a completely new and special perspective. It is becoming clear that national parks will play a crucial role in the future of tourism. Ensuring long-term sustainable management and development is of primary importance. This is necessary in order not only to preserve the value for tourism and society as a whole, but also to increase it.[40] [41] [42] [43]

In summary, national parks play a crucial role in modern tourism. As a result, they have become essential tourism destinations, offering recreation, education and adventure in unique natural environments. They meet the growing demand for sustainable travel experiences and represent the principles of a nature-based form of tourism. Finally, it can be said that "[without] national parks (…) many of these regions would be far less or not at all frequented by tourists ."[44]

---

[39](Berzina, Livina, 2008, p. 100)
[40]See (Berzina, Livina, 2008)
[41]See (National park brings opportunities for nature, business and tourism , n.d.)
[42]See (Revermann, Petermann, 2002)
[43]See (Tschurtschenthaler, 2000)
[44](Tschurtschenthaler, 2000, p. 88)

Now that the ever-increasing importance of national parks for tourism has been discussed, it is also necessary to understand the economic aspects in more detail. It is undeniable that tourism in national parks has a significant economic impact. But how exactly does national park tourism contribute to the economy and do residents fundamentally benefit from it? This question will be addressed in the next section.

2.2.2 National park tourism as an economic engine

Tourism in national parks is not only an important factor in the preservation of the parks themselves, but it also plays an important role as an economic engine. From direct spending by visitors to creating jobs and supporting local businesses, the economic benefits of national park tourism are diverse. In this section, the economic aspects are analyzed in more detail. Conclusions can be drawn from this as to whether and how tourism in national parks contributes to strengthening the economy.[45]

National parks have an extraordinary impact on local economies in many ways. A distinction is made between the expenses of tourists and the national park-related expenses of the government. If you look at the former, money is spent in many parts of the trip, which benefits the local community. What is particularly crucial is that for every dollar spent in the national park, $10 flows back into the economy. Parts of their expenses include not only entrance fees, but also costs for overnight accommodation and other tourist activities, rental car prices, taxes, expenses in restaurants and souvenirs. These can be divided into three effects:[46 47 48 49 50]

- direct impact: regarding local services and products

- indirect effects: Companies and service areas that have the visitors as customers but obtain their services and products from other local companies

- Induced impacts: The expenditure of the income generated directly or indirectly by tourism, which in turn stimulates further economic activities

"It should be kept in mind that the economic impacts do not describe the total value of the national park as they only reflect a part of the use."[51] Building on the knowledge that the tourism industry has close interrelationships with various economic sectors, a closer look is crucial as we move forward.[52 53]

---

[45]See (Martinez, 2003)
[46]See (Fish, 2009)
[47]See (Land, 2019)
[48]See (Martinez, 2003)
[49]See (Huhtala et al., 2010)
[50]See (Mika et al., 2015)
[51](Huhtala et al., p. 7)
[52]See (Revermann, Petermann, 2002)
[53]See (Tschurtschenthaler, 2000)

"[National parks] also contribute to local economies through employment of agency personnel; park operations and capital expenditures; by influencing park-related employment and economic development, especially in amenity and tourism support industries; and through associated local household spending."[54] But a direct connection can also be seen on a national level. Economists from the National Park Service showed particular interest. Lynne Koontz, Catherine C. Thomas, and Egan Cornachione conducted peer-reviewed research into visit expenditures. The results show that 318,000,000 national park visitors spent around $20.3 billion in towns within a radius of almost 100 kilometers from the park. [55]This effect consequently led to the support of almost 330,000 jobs nationwide, of which "(…) more than 268,000 jobs were found in these gateway communities, and had a cumulative benefit to the US economy of $40.1 billion." The interesting question now [56]is in which segment the national park visitors have the largest expenses and what dependence there is as a result. The investigation is based on the most current values for 2022, which were published by the National Park Service. The table below summarizes the results:[57]

Table 1: Economic effects of national park visitors in America 2022

Number of visitors to the national park	312,000,000
Expenses in surrounding communities	$23,900,000
Number of jobs supported	378,000
Amount of employment income	$17,500,000
Amount of added value	$29,000,000
Amount of economic output in the national economy	$50,300,000

Source: own representation based on (NPS.gov, 2023i)

If you take a detailed look at the current data, you can see that although 6 million fewer travelers visited national parks in the United States, they had a greater economic impact. Spending in surrounding communities increased by $3,600,000 and nearly 50,000 additional jobs were supported. [58]Looking at the share of economic output at the national level, there was an increase of 25.44% [59]within three years. If you look at

---

[54](Fish, 2009, p.1)
[55]See (Land, 2019)
[56](Country, 2019)
[57]See (NPS.gov, 2023i)
[58]See (Land, 2019)
[59]Previously the number was 40,100,000, 3 years later it was 50,300,000. The difference of $10,200,000 is 25.44% rounded

the expenditure in the surrounding communities, you can see that most of the money, at a percentage of 37.63%, went into overnight accommodation, followed by 19.32% into catering. In third place was spending at gas stations at 10.97%, closely followed by the leisure and entertainment industry at just under two percent. There were similar expenses in retail. At the bottom were expenses for transport (6.7%), purchases (6.46%) and ultimately costs for camping in the national park (2.15%). These results are illustrated in Figure 3.[60] [61]

Figure 3: Segments and their shares regarding expenditure

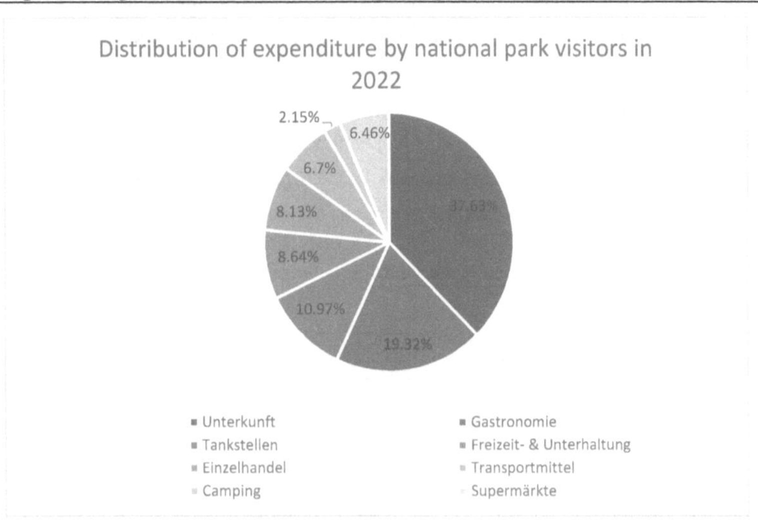

Source: own illustration based on (NPS.gov, 2023i)

The direct connection between visitors to the national park and the positive economic impact on a regional and national level is visible. It is extremely remarkable that the people who decide to go on a trip to a national park play a crucial role for various economic enterprises. The greater the demand and the more travelers visit the region, the higher the chance that a new company can make a name for itself in the area. An optimal basis is also offered for local residents who would like to reorient themselves professionally. In addition, the expenditures of the above-mentioned economic sectors contribute directly to tax revenue, which in turn has a positive effect on overall national economic performance. [62] [63] [64] [65]

---

[60]See (NPS.gov, 2023i)
[61]See (Land, 2019)
[62]See (National Governors Association, undated)
[63]See (NPS.gov,2023i)
[64]See (National Governors Association, undated)
[65]See (Budeanu, 2005)

In conclusion, the effects of national park tourism as a catalyst for the local economy can be assessed at national, regional and local levels. National park tourism not only contributes to improving social benefits, but also to the preservation of cultural heritage and the landscape. Likewise, the associated economic benefits have an immense impact and do not just include a positive increase in income. The growing demand in tourism affects tax revenues with a direct link to the gross national product. Furthermore, the overall economic effect has a notable influence on the local population. The income and jobs generated by tourism can significantly improve the quality of life of the population and create new professional opportunities. In addition, the local population benefits from expanded leisure activities, as the increasing number of visitors leads to the development of restaurants, cafés and tourist attractions. Ultimately, tourism in national parks provides both economic and socioeconomic opportunities for the sustainable development of communities.[66] [67] [68]

By specifically evaluating the economic effects of visitors to the national park, it can be made possible to facilitate management and planning decisions at regional and local levels.[69] [70]

The findings of the chapter serve as an optimal basis for the third part of the case study, which also analyzes the economic aspects of tourism in Joshua Tree National Park. This serves to draw conclusions as to whether tourism in Joshua Tree National Park represents a benefit - with regard to the economy and the quality of life of the residents.

---

[66]See (Livina, Berzina, 2008)
[67]See (Szabo, Ujhelyi, 2023)
[68]See (Hardner, McKenney, 2006)
[69]See (Martinez, 2003)
[70]See (Fish, 2009)

List of references

Arbter, K. (2011). PUBLIC PARTICIPATION STANDARDS: Practice Guide. *Federal Ministry of Agriculture, Forestry, Environment and Water Management* , Federal Chancellery of Austria.

Becker, A. (2022). 4 Methodical approach. In *De Gruyter eBooks* (pp. 91–104). DeGruyter. https://doi.org/10.1515/9783110770773-004

Berzina, I. & Livina, A. (2008). The Model on Estimating Economic Benefit of Nature-based Tourism Services of Territories of National Parks, Latvia. *DestiNet.eu* . https://destinet.eu/resources/. . .-various-target-groups/model-estimating-economic-benefit-nature-based-tourism/download

BMZ. (2022, May 25). *Tourism* . Federal Ministry for Economic Cooperation and Development. Retrieved on October 21, 2023, from https://www.bmz.de/de/themen/tourismus#:~:text=With%20dem%20Anstieg%20der%20Number , direct%20connection%20mit%20dem% 20Tourism .

Bramwell, B. & Lane, B. (2011). Critical research on the governance of tourism and sustainability. *Journal of Sustainable Tourism* , *19* (4–5), 411–421. https://doi.org/10.1080/09669582.2011.580586

Budeanu, A. (2005). Impacts and responsibilities for sustainable tourism: a tour operator's perspective. *Journal of Cleaner Production* , *13* (2), 89–97. https://doi.org/10.1016/j.jclepro.2003.12.024

Choi, F. & Marlowe, T. (2012). The Value of America's Greatest Idea: Framework for Total Economic Valuation of National Park Service Operations and Assets. *Harvard Kennedy School of Government* . https://www.nps.gov/subjects/policy/upload/Task-4-Joshua-Tree-Case-Study-The-Value-of-America-s-Greatest-Idea-Choi-and-Marlowe-2012. pdf

German Society for International Cooperation (GIZ) GmbH. (2020). The tourism value chain: Analysis and application approaches for development cooperation projects. *GIZ* . https://www.giz.de/de/downloads/giz2020-de-wertschoepfungskette-tourismus.pdf

Ennew, CT (2014). Understanding the economic impact of tourism. *Warwick* .
https://www.academia.edu/1368653/Understanding_the_economic_impact_of_t
ourism

Eyrich, T. (2023). *The American dream at the edge of the earth* . UC Riverside.
Retrieved October 9, 2023, from https://news.ucr.edu/ucr-magazine/spring-
2023/the-american-dream-at-edge-the-of-the-earth

Fish, T.E. (2009). Assessing economic impacts of national parks. *National Park
Service* , *Volume 26* . http://www.cesu.psu.edu/materials/parksci7-
economicimpacts.pdf

Hardner, J. & McKenney, B. (2006). The US National Park System: An economic asset
at risk. *Hardner & Gullison* .
https://npca.s3.amazonaws.com/documents/1109/1500b64a-ed91-4403-95fb-
018d662be838.pdf?1445978502

Huhtala, M., Kajala, L., Vatanen, E. & Fi, WM (2010). Local economic impacts of
national park visitors' spending in Finland: The development process of an
estimation method. *ResearchGate* .
https://www.researchgate.net/publication/228455705_Local_economic_impacts
_of_national_park_visitors'_spending_in_Finland_The_development_pro
cess_of_an_estimation_method

Klapf, S. (2005). *Gentle tourism in the national park - an opportunity to increase
acceptance?* [Diploma thesis]. JOHANNES KEPLER UNIVERSITY OF LINZ.
https://www.parcs.at/npg/pdf_public/2019/12762_20191211_134533_Klapf200
5-SanfterTourismusimNationalpark.pdf

Kumar, J. & Hussain, K. (2014). Evaluating Tourism's Economic Effects: Comparison
of different approaches. *Procedia - Social and Behavioral Sciences* , *144* , 360–
365. https://doi.org/10.1016/j.sbspro.2014.07.305

Martinez, L. (2003). National Treasures as Economic Engines The Economic Impact of
Visitor Spending in California's National Parks. *NATIONAL PARKS
CONSERVATION ASSOCIATION* .
http://www.sequoiaforestkeeper.org/images/CA%20NP%20economic%20benef

its%20NPCA%202009.pdf

Mika, M., Zawilinska, B. & Pawlusinski, R. (2015). Exploring the economic impact of national parks on the local economy. Functional approach in the context of Poland's transition economy. *Human Geographies – Journal of Studies and Research in Human Geography* , *10* (1). http://humangeographies.org.ro/articles/101/a_10_1_mika.pdf

Mojave Desert Land Trust. (2018). ECONOMIC IMPACT OF DESIGNATION OF NATIONAL MONUMENTS IN SAN BERNARDINO COUNTY: Sand to Snow, Castle Mountains and Mojave Trails National Monuments. *NPS history* . http://npshistory.com/publications/blm/mojave-trails/economic-impacts-2018.pdf

National Governors Association. (n.d.). https://www.nga.org/wp-content/uploads/2020/06/GAO-Act-Economic-Benefits-Factsheet.pdf. In *NGA.org* . Retrieved September 16, 2023, from https://www.nga.org/wp-content/uploads/2020/06/GAO-Act-Economic-Benefits-Factsheet.pdf

National Parks Association of NSW. (2023, May 2). *Why are national parks important?* https://npansw.org.au/what-we-do/why-are-national-parks-important/

National park brings opportunities for nature, business and tourism . (n.d.). Baden-Württemberg.de. Retrieved on October 14, 2023, from https://www.baden-wuerttemberg.de/de/service/alle-meldeen/melde/pid/dreimal-mehr-opportunities-natur-wirtschaft-und-tourismus#:~:text=The%20report%20shows%2C%20that%20the%20people%20in%20the%20region

NPS.gov. (2023a). *Stats Report Viewer* . National Park Service. Retrieved October 14, 2023, from https://irma.nps.gov/Stats/SSRSReports/Park%20Specific%20Reports/Traffic%20Counts?Park=JOTR

NPS.gov. (2023b). *Stats Report Viewer* . National Park Service. Retrieved October 4, 2023, from https://irma.nps.gov/Stats/SSRSReports/Park%20Specific%20Reports/Visitation%20by%20Month?Park=JOTR

Nyirarwasa, A., Han, F., Pan, X., Mind'je, R., Maniraho, AP, Gasirabo, A., Udahogora, M., Mtewele, ZF & Umwali, ED (2020). Evaluating the relationship between national park management and local communities' perceptions based on survey, a case of Nyungwe National Park, Rwanda. *Journal of Geoscience and Environment Protection* . https://doi.org/10.4236/gep.2020.812007

Revermann, Chr. & Petermann, Th. (2002). Tourism in large protected areas - interactions and opportunities for cooperation between nature conservation and regional tourism. Final report. *Office for Technology Assessment at the German Bundestag (TAB)* , TAB work report; 77. https://doi.org/10.5445/ir/1000103520

Schmude, J. & Bartl, E. (2023). *Tourism industry* . springerprofessional.de. Retrieved on September 21, 2023, from https://www.springerprofessional.de/tourismuswirtschaft/24601666

Specht, G., Santos, A.D. & Bingemer, S. (2004). The case study in the cognitive process: the case study method in economics. In *German University Press eBooks* (pp. 539–563). https://doi.org/10.1007/978-3-322-81694-8_24

Stevens, TH, More, TA & Markowski - Lindsay, M. (2014). Declining national park visitation. *Journal of Leisure Research* , *46* (2), 153–164. https://doi.org/10.1080/00222216.2014.11950317

Stynes, D.J. (1997). Economic Impacts of Tourism. *PSU.edu* .

Stynes, D.J. (2003). Economic Impacts of National Park Visitor Spending on Gateway Communities: Systemwide Estimates for 2001. *NPS history* , Michigan State University. http://npshistory.com/publications/social-science/mgm2/economic-impacts-2001.pdf

Szabo, A. & Ujhelyi, G. (2023). National Parks and Economic Development. *uh.edu* . https://uh.edu/~gujhelyi/nps.pdf

Thomas, C. C., Huber, C. & Koontz, L. (2015). *2014 National Park Visitor Spending Effects: Economic contributions to local communities, states, and the nation* . National Park Service. Retrieved September 12, 2023, from https://pubs.usgs.gov/publication/70148496

Tschurtschenthaler, P. (2000). National park and tourism. In *A Tourism Economic Consideration* (Vol. 11, pp. 77–124). Zoological-botanical database. https://www.zobodat.at/pdf/Natur-in-Tirol_11_0077-0124.pdf

Vellas, F. (2011). *The Indirect Impact of Tourism: An Economic Analysis* . Third Meeting of T20 Tourism Ministers. Paris, France. Retrieved September 14, 2023, from https://webunwto.s3.amazonaws.com/imported_images/28700/111020-rapport_vellas_en.pdf

By Rimscha, MB & Sommer, C. (2014). Case studies in communication studies. In *Springer eBooks* (pp. 1–13). Springer. https://doi.org/10.1007/978-3-658-05723-7_23-1

Zaei, M.E. & Zaei, M.E. (2013). THE IMPACTS OF TOURISM INDUSTRY ON HOST COMMUNITY. *European Journal of Tourism Hospitality and Research* , *1* (2), 12–21. https://www.researchgate.net/profile/Mansour-Esmaeil-Zaei/publication/357736048_The_Impacts_of_Tourism_Industry_on_Host_Co mmunity/links/61dd3219323a2268f997f90d/The-Impacts-of-Tourism-Industry-on-Host-Community.pdf